SUN

In Praise of the Golden Eye of Heaven

IN PRAISE OF THE GOLDEN EYE OF HEAVEN

SUN

EDITED BY

Paula Rees

HarperSanFrancisco

An Imprint of HarperCollins*Publishers*

Permissions acknowledgments begin on page 84
and are a continuation of this copyright page.

HarperSanFrancisco
1160 Battery Street
San Francisco, CA 94111

FIRST EDITION
Library of Congress Cataloging-in-Publication Data
Sun: In Praise of the Golden Eye of Heaven / edited by Paula Rees.—1st ed.
p. cm.
Includes bibliographical references and index.
ISBN 0-06-251261-7 (pbk.: acid free paper)
1. Sun—Literary collections I. Rees, Paula.
PN6071.S88S86   1996
808.8'036—dc20   95-43447 CIP
96 97 98 99 00 HAD 10 9 8 7 6 5 4 3 2 1

For the sun is not inappropriately called by some people

the lantern of the universe, its mind by others,

and its ruler by still others.

NICOLAUS COPERNICUS

CONTENTS

SUN

Before the measurement of time or territories, man was bound to the sun. Our star, center, and source of energy, the sun influences us down to the cellular level. Hailed as our father, provider, creator, and even god, the sun's emergent and life-providing energy commands our respect.

The staccato of today's rapid change shakes our belief in time as a linear event. In our confusion, the sun gently awakens our memory and helps us contemplate a more fluid notion of time and space. A focus on natural cycles, simpler rhythms and pacing returns as we prayerfully pass the cusp of another millennium. We are in awe of the sun's cyclic continuity. The sun is a life-giver, providing hope for renewal and growth.

In our daily lives we have been "sheltered" from the sun. Sensing this, we seek its companionship and nourishment in leisure and vacation time, which is frequently designed to reconnect us with its vital warmth. Moving out of our temperature-controlled, artificially lit, filtered environments, we put on sunglasses and join in the larger ecliptic dance.

Color is the sun's messenger. From this golden disk, the full spectrum of light is revealed. The sun illuminates all. Its appearance gives us solace, as it gently wakens us with rosy-fingered dawn; we encounter light and movement, gaining perspective through contrast and shadow. The sun's personality is understood by reflection; we would be blinded for meeting its gaze.

Once, we believed we were at the center of the universe, only to learn to the contrary—that we revolve around the sun. We have always turned intuitively to the sun to understand our sense of place and power. Definitions of earthly power cloak themselves in solar imagery. We are caught in a moment of insignificance and amazement when we learn that this great fire wheel is only one of one hundred million stars in our small corner of the universe. But we forget this fact, we regress, taking ourselves very seriously and feeling intensely important. The sun lightly *"answers all our questions with its flaming laughter."*

Paula Rees

Signifying the Hand of God

And God said, Let there be light: and there was light.
And God saw the light, that it was good:
and God divided the light from the darkness. . . .

And God said, Let there be lights in the firmament of the
heaven to divide the day from the night;
and let them be for signs, and for seasons, and for days, and years:
And let them be for lights in the firmament of the heaven to
give light upon the earth: and it was so.

GENESIS I: 3-4, 14-15

The mind is turned toward God in the same way
that the eye is directed toward the light of the sun;
next it sees the light of the sun; third, in the light of
the sun it perceives the colors and shapes of things.
Therefore the eye, at first dark and, like Chaos,
formless, loves the light while it looks toward it;
in so looking, it is illuminated; in receiving the ray,
it is informed with the colors and shapes of things.

MARSILIO FICINO

The heavens declare God's grandeur
> and the radiance from which they arise.
Each dawn tells of his beauty;
> each night shines with his grace.
Their testimony speaks to the whole world
> and reaches to the ends of the earth.
In them is a path for the sun,
> who steps forth handsome as a bridegroom
> and rejoices like an athlete as he runs.
He starts at one end of the heavens
> and circles to the other end,
> and nothing can hide from his heat.

PSALM 19

translation by Stephen Mitchell

Ancient sun, eternally young,
giver of life and source of energy,
    In coal and oil, in plant and wind and tide,
    in spiritual light and human embrace,
You kindle the heavens, you shine within us
(for we are suns with hearts afire—
    we light the world as you light the sky
    and find clouds within whose shadows are dark);
We give thanks for your rays . . . .

CONGREGATION OF ABRAXAS

11

The herds are at peace in their pastures,
    trees and the vegetation grow green;
Birds start from their nests,
    wings wide spread to worship your Person;
Small beasts frisk and gambol, and all
    who mount into flight or settle to rest
       live, once you have shone upon them;
Ships float downstream or sail for the south,
    each path lies open because of your rising;
Fish in the River leap in your sight,
    and your rays strike deep in the Great Green Sea.

It is you create the new creature in Woman,
    shape the life-giving drops into Man,
Foster the son in the womb of his mother,
    soothe him, ending his tears;
Nurse through the long generations of women
                to those given Air,
    you ensure that your handiwork prosper.
When the new one descends from the womb
    to draw breath the day of his birth,
You open his mouth, you shape his nature,
    and you supply all his necessities.

PHARAOH AKHENATEN
Hymn to the Sun

Now this day, my Sun Father,
Now that you have come out standing
To your sacred place.

That from which we draw the water of life
Prayer meal—
Here I give unto you.

Your long life,
Your old age,
Your waters,
Your seeds,
Your riches,
Your power,
Your strong spirit.
Of all these, to me may you grant.

EGYPTIAN PRAYER TO SUN GOD

He who inhabits the sun, yet is within the sun,
whom the sun does not know, whose body the sun is,   15
and who controls the sun from within—
He is your Self, the Inner Controller, the Immortal.

BRIHADARANYAKA UPANISHAD, III. vii. 9

PROVERB · *The sun*

shines upon

all alike.

Through all thy creatures Lord be praised
Especially through our Brother Sun
Who brings us day and illumines with Thy light
And he is gloriously radiant and splendid
Bespeaking Thee to us, Most High

Praise be to Thee Lord for Sister Moon and the stars
In the heavens Thou hast set them clear and precious and fair.

SAINT FRANCIS OF ASSISI

O King of the sun's brightness,
who alone has knowledge of our purpose,
be with us every day,
be with us every night,
be with us every night and day,
be with us every day and night.

CELTIC PRAYER

17

The sun at home warms better

than the sun abroad.

"When the Sun rises, do you not see a round disk of fire
    somewhat like a Guinea?"
O no, no, I see an innumerable company of the
    Heavenly host crying   19
"Holy, Holy, Holy is the Lord God Almighty."

WILLIAM BLAKE

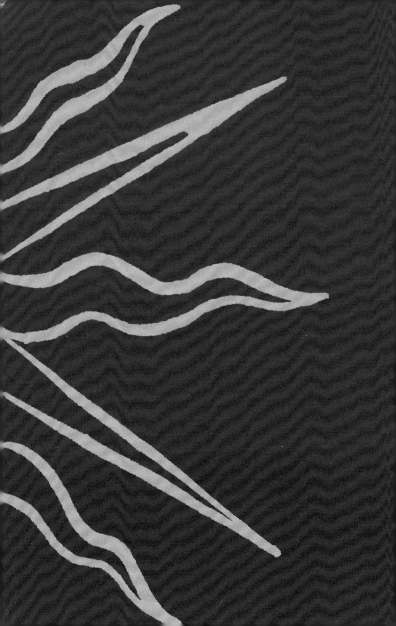

It Has Become Light

The sun is

the poor man's

blanket.

The breeze at dawn has secrets to tell you.
>   Don't go back to sleep.
You must ask for what you really want.
>   Don't go back to sleep.
People are going back and forth across the doorsill
>   where the two worlds touch.
The door is round and open.
>   Don't go back to sleep.

23

RUMI

The sun does not rise

for one man alone.

The Sun, the Light of the world,
I hear Him coming.
I see His face as He comes.
He makes the beings on earth happy,
And they rejoice.
O Wakan-Tanka, I offer to You this world of Light.

OGLALA SIOUX
Sacred Song of the Sun Dance

25

To him whose elastic
and vigorous thought
keeps pace with the sun,
the day is a perpetual morning.

HENRY DAVID THOREAU

Standing by my bed

In gold sandals
Dawn that very
moment awoke me

Hesperus, you herd
home whatever
Dawn's light dispersed

You herd sheep—herd
goats—herd children
home to their mothers.

SAPPHO OF LESBOS
Standing By My Bed

He maketh his sun to rise on the evil and the good,

and sendeth the rain on the just and unjust.

MATTHEW V:45

Busy old fool, unruly Sun,
    Why dost thou thus,
Through windows and through curtains call on us?
Must to thy motions lovers' seasons run?
    Saucy pedantic wretch, go chide
    Late school-boys, and sour 'prentices,
Go tell court-huntsmen that the King will ride,
Call country ants to harvest offices;
Love, all alike, no season knows, nor clime,
Nor hours, days, months, which are the rags of time.

29

JOHN DONNE
The Sun Rising

Sunrise is the time that you will be able to find out how to help somebody close to you who you think needs help even if he doesn't think so. At sunrise everything is luminous but not clear.

NORMAN MACLEAN
A River Runs Through It

Not afraid of anything is he,
    and then goes cowering forth, tread paced to meet an obstacle
at every step. Consistent with the
        formula—warm blood, no gills, two pairs of hands and a few
            hairs—that

is a mammal; there he sits in his own habitat,
    serge-clad, strong-shod. The prey of fear, he, always
        curtailed, extinguished, thwarted by the dusk, work
            partly done,

says to the alternating blaze,
    "Again the sun!
      anew each day; and new and new and new,
      that comes into and steadies my soul."

31

MARIANNE MOORE
The Pangolin

The sun has climbed the hill, the day is on the downward slope.
Between the morning and the afternoon, stand I here with my soul,
      and lift it up.
My soul is heavy with sunshine, and steeped with strength.
The sunbeams have filled me like a honeycomb,   33
It is the moment of fulness,
And the top of the morning.

D. H. LAWRENCE
The Mid-Day Verse

He is very

blind who

cannot see

the sun.

What, hast thou run the Race? Art going down?
Thou seemest angry, why dost on us frown?
Yea wrap thy head with Clouds, and hide thy face,
As threatening to withdraw from us thy Grace?
Oh leave us not! When once thou hid'st thy head,
Our Horizon with darkness will be spread,
Tell's who hath thee offended? Turn again:
Alas! too late Entreaties are in vain!

JOHN BUNYON
Of the Going Down of the Sun

Blazing in Gold and quenching in Purple
Leaping like Leopards in the Sky . . .
And the Juggler of Day is gone.

EMILY DICKINSON
Who Is the East?

A land not mine, still
forever memorable,
the waters of its ocean
chill and fresh.

Sand on the bottom whiter than chalk,
and the air drunk, like wine,
late sun lays bare
the rosy limbs of the pine trees.

Sunset in the ethereal waves:
I cannot tell if the day
is ending, or the world, or if
the secret of secrets is inside me again.

ANNA AKHMATOVA

The sun loves to peer

into a home where love is.

Venus, the mother of that bastard Love
Which doth usurp the world's great marshal's name,
Just with the sun her dainty feet doth move,
And unto him doth all her gestures frame:
Now after, now afore, the flatt'ring dame
With divers cunning passages doth err,
Still him respecting that respects not her.

For that brave Sun, father of the day,
Doth love this Earth, the mother of the night;
And, like a reveller in rich array,
Doth dance his galliard in his leman's sight,
Both back and forth and sideways passing light.
His gallant grace doth so the gods amaze
That all stand still and at his beauty gaze.

SIR JOHN DAVIES
Orchestra

Giver of Life

The sun is the king

of torches.

I say Live, Live because of the sun,
The dream, the excitable gift.

43

ANNE SEXTON
Live

44   For the sun, together with the ambient, is always in
some way affecting everything on the earth, not only by the
changes that accompany the seasons of the year to bring about
the generation of animals, the productiveness of plants,
the flowing of waters, and the changes of bodies, but also by its
daily revolutions furnishing heat, moisture, dryness, and cold in
regular order and in correspondence with its positions relative
to the zenith.

CLAUDIUS PTOLEMY
Tetrabiblos

Life, one might say, is the daughter of the sun.
The rays whose waves fall upon the chlorophyll in the
leaves of plants, produce a chemical action which man
has been unable to obtain by the synthetic methods
which are now at the service of science. The living
energy of the sun is absorbed and transformed: his
potential energy goes to sleep, if it is permissible so to
express it, in the leaves, in the seeds of the plants,
in the albuminoid substances which are produced in
vegetable cells. . . .

A. MOSSO

45

In the east is the dwelling of the sun.
On top of this dwelling place
The sun comes up and travels over our heads.

Below we travel.
I raise my right hand to the sun
And then stroke my body
In the ceremonial manner.

PAPAGO
Ceremonial Sun Song

· *In every*
·
· *country the*
·
· *sun rises in*
·
· *the morning.*

Ƴ Ο ୧

We are glad to be here.

We are going to give our food to this our country.

Make us strong; make us lucky.

Keep the women in good health so that they may dig camas
        and other roots in plenty for the winter.

Sun, you give us what we want so that we may be happy.

Sun, you are all-knowing. You want us to have a good time
        and to live well.

Now, great sun, you know everything because you are over us.

MODOC

Prayer Before the First Meal in a New Summer Camp

Humanity was a pulse in the life of the Sun, which lay burning there in a vast white explosion of varying kinds of light, or sound, some stronger and thicker, some tenuous, but at all forces and strengths, which fluid lapped out into space holding all these crumbs and drops and little flames in a dance—and the force that held them there, circling and whirling in their dance, was the Sun, the energy of the Sun, and that was the controlling governor of them all, beside whose strength, all the subsidiary laws and necessities were nothing. The ground and soul and heart and centre of this little solar system was the light and pulse and song of the Sun, the Sun was King.

DORIS LESSING
Briefing for a Descent into Hell

Who dares declare his love to the
sun, who can have any relationship with the sun . . . it is too
enormous and overwhelming, we cannot even look at it with
our eyes open without being immediately blinded,

the sun is the master of space, the
ruler of our universe, the roaring lion in the high which is at
the same time the deep, it must be utterly inconsequential to
the sun whether our small planet exists or not, the sun is too
immense to share any of our troubles,

the existence of the sun could be
imagined as an unending suffering rather than as some kind of
bliss, it does tear itself apart constantly in a rage that never
solidifies, it is an expression of violent wrath rather than of
lovingness,

the sun is so far from everything
human that it seems like a bleeding taunt that human
existence depends on it, what are we with our earth but red
or blue lice in the colorless darkness of the heavenly blanket,
if we burned in a moment there would be no more of a
crackling than if a louse were crushed,

the sun is one single flaming
rebellion, the truly ongoing revolution, the ultimate law of
self-destruction,

the sun has also been worshipped
as a god and perhaps it is the greatest fathomable or unfath-
omable symbol of god, a god who creates and destroys himself without pause while eternity turns around itself, the sun
has certainly sacrificed no son to save us, it answers all our
questions only with its flaming laughter,

how can we imagine any other
to the sun than this our blind and circumstantial
dependency,

the sun owes us nothing and we
cannot expect anything from it, it does not know what it has
created, we are the abandoned children of coincidence in the
unending, meaningless emptiness.

ARTUR LUNDKVIST
Journeys in Dream and Imagination

I look up. The book is now in shadow. The earth has moved. Afternoon sunlight now enters the southern windows and slants across the floor. Sunlight, slanting across a windowsill like this, or resting upon a wall, has sometimes rescued me (when I noticed it) from despair. Does the sun know what it's doing? Could that which has lifted us up out of the chaos be less intelligent, less conscious than we are? Is all this towering largesse—even prodigality—only a manifestation of automatic processes? Or do we suffer from a kind of chauvinism of scale, which blinds us to the presence of awareness in any creature much greater or smaller, much longer—or shorter-lived than ourselves?

We owe our very lives to the sun. Nothing in the facts of science contradicts this assertion. How is it, then, that we feel no gratitude?

53

MARTHA HEYNEMAN
Fields of Force

He who avoids the sun

will always be cold.

Lantern of the
Universe

The sun

can be seen by

nothing but its

own light.

In this most beautiful temple, who would place this lamp in another or better position than that from which it can light up everything at the same time? For the sun is not inappropriately called by some people the lantern of the universe, its mind by others, and its ruler by still others.

NICOLAUS COPERNICUS

What if the Sun
Be Center to the World, and . . .

58

The Planet Earth, so steadfast though she seem,
Insensibly three different Motions move?

JOHN MILTON
Paradise Lost

The constellations, along with the nights, steal
away like thieves, making way for the sun who
gazes on everyone.

The rays that are his banners have become visible
from the distance, shining over mankind like
blazing fires.

Crossing space, you are the maker of light,
seen by everyone, O sun. You illumine the whole,
wide realm of space.

You rise up facing all the groups of gods,
facing mankind, facing everyone, so that they
can see the sunlight. . . .

You cross heaven and the vast realm of space,
O sun, measuring days by nights, looking upon
the generations. . . .

We have come up out of darkness, seeing the
higher light around us, going to the sun, the god
among gods, the highest light.

RIG VEDA

The sun! The sun! And all we can become!
And the time ripe for running to the moon!
In the long fields I leave my father's eye;
And shake the secret from my deepest bones;
My spirit rises with the rising wind.

THEODORE ROETHKE
What Can I Tell My Bones?

The sun is a glorious creature, and its beams extend to
the utmost stars; by shining on them it clothes them with
light, and by its rays exciteth all their influences.
It enlightens the eyes of all the creatures: it shineth on
forty kingdoms at the same time, on seas and continents
in a general manner; yet so particularly regardeth all, that
every mote in the air, every grain of dust, every spire of
glass is wholly illuminated thereby as if it did entirely
shine upon that alone. Nor does it only illuminate all
these objects in an idle manner; its beams are operative,
enter in, fill the pores of things with spirits, and
impregnates them with powers, cause all their emanations,
odors, virtues, and operations; springs, rivers, minerals
and vegetables are all perfected by the sun; all the motion,
life and senses of birds, beasts and fishes dependeth on
the same. Yet the sun is but a little spark among all the
creatures that are made for the Soul; the Soul being
the most high and noble of all, is capable of far higher
perfections, far more full of life and vigour in its uses.
The sphere of its activity is illimited, its energy is
endless upon all objects. It can exceed the heavens in its
operations, and run out into infinite spaces. Such is the
extent of knowledge that it seemeth to be the Light of
all Eternity.

61

THOMAS TRAHERNE

Light in itself is something akin to the soul. . . .
and so it is that consonant that the solar body,
wherein the light is present as in its source,
is endowed with a soul which is the originator,
the preserver, and the continuator. And the function
of the sun in the world seems to persuade us of
nothing else except that just as it has illuminated all
things, so it is possessed of light in its body; and as
it has to make all things warm, it is possessed of
heat; as it has to make all things live, of a bodily
life; and as it has to move all things, it itself is the
beginning of the movement; and so it has a soul.

JOHANNES KEPLER

63

No sun shines

without some cloud.

The sun shines

for all the

world.

Love then came, as it desired,
and as under the appearance of a cloud
luminously it swooped down on me;
completely on my head
I saw it settle;
and it made me cry out,
for I was in terror.
Nevertheless, after having then flown away,
Love left me alone,
and while I arduously searched for it
then suddenly, completely
it was in me in a conscious manner,
in the center of my heart;
like a truly heavenly body,
I saw it like the solar disc . . .

ST. SYMEON
Hymn 17

Make use of

the sun while

it shines.

Under the new-made clouds and happy as the heart was long,
In the sun born over and over,
 I ran my heedless ways.

67

DYLAN THOMAS
Fern Hill

The sun shines

even into a little room.

We stand amid time beginningless and endless,
    we stand amid evil and good,
All swings around us, there is as much darkness as light,
The very sun swings itself and its system of planets around us,
Its sun, and its again, all swing around us.

69

WALT WHITMAN
With Antecedents

101

SUN CALENDAR

JANUARY 1   *New Year's Day,*
first day of the solar calendar

JANUARY 4   *Earth perihelion,*
day the earth is closest to the sun

FEBRUARY 2   *Imbolic or Candlemas Day,*
cross-quarter day marking the halfway point
between Winter Solstice & Vernal Equinox

MARCH 21   *Vernal Equinox,*
the beginning of spring,
with equal hours of light and dark

MAY 1   *Beltane, May Day festival marking*
the cross-quarter day halfway between
Spring Equinox & Summer Solstice

MAY 2   *Fire Festival,*
ancient Roman holiday of the Sun

JUNE 21 *Summer Solstice,*
longest day of sunlight

JULY 5 *Earth aphelion,*
day the earth is farthest from the Sun

JULY 17 *Festival of Amaterasu-o-Mi-Kami,*
Japanese festival of the female sun goddess

AUGUST 1 *Lammas, cross-quarter day marking*
the halfway point between
Summer Solstice & Autumnal Equinox

SEPTEMBER 22 *Autumnal Equinox,*
beginning of autumn, with equal hours
of light and dark

OCTOBER 31 *Halloween, All Hallows' Eve, Samhain,*
cross-quarter day between
Autumnal Equinox & Winter Solstice

DECEMBER 21 *Winter Solstice, shortest day of sunlight*

e sata satarapsi aayyah So
uya nap soleil Sonce sa
ne watà Zon schem
yyah sun sol günes i'lios
ol matahari sinne Sol
satarapsi jua suno sūraj
a soleil diell Sun a
innakara nap soleil Sonce
ata bhānu apsumati S
aayyah matahari Sur
ko sun suriya diell mat
suno grian satarapsi
günes ravi shams go
jua hari aurinko sun So

Sun Deities

zon aurinko suria
sataransi jua suno suraj
ch grian surya sol
shams sol jua su
dzuh sinne matahari s
sata soleil sol
inko sure bhānu haru
ā aayyah i'lios soleil
l stonce sonce sata jua s
jua suno nap soleil ma
ri sinne divākara b
zuh dinnakara aayyah satas
ansumali sātaransi aay
aayyah sol ansumali sata

AMATERASU

The "shining from heaven" sun-goddess of Japanese Shintoism, venerated in the shrine at Ise as the divine progenitor of the Japanese imperial family. Her epithet is Omikami: "great and exalted divinity." She wove the robes of the gods.

APOLLO

Greek god, son of Zeus, and twin brother to Artemis, goddess of hunting and the moon. Most frequently identified as the god of prophesy, philosophy, music, and medicine. He was wise, young, strong, and handsome. As Phoebus Apollo, "the shining one," he was also associated with Helios, the sun-god.

ARINNA

As the Hittite "Queen of Heaven and Earth," this sun-goddess protected the kingdom and assisted in its wars. Her cult symbol is the sun disc and is often identified with the Hurian goddess of heaven, Hebat.

ATON/ATUM

In Ancient Egypt, primarily the designation of the visible disc of the sun, which was regarded as an evening aspect of Re. In the New Kingdom, the sun-disc was personified, and under King Amenophis IV was declared to be the one true god. Portrayals from this period show the sun-disc, whose rays are arms which end in hands, bearing the symbol of life—the crux ansata.

BASTET (BAST)

Solar goddess depicted as a woman with a cat's head. A popular Egyptian deity of pleasure.

BEHDETITE

This god, taking the form of a crouching falcon, was venerated in the Egyptian town of Behtet. Very early on he became a local form of the great falcon-god Horus. The proper symbol for the solar deity is the disc of the sun, fitted with a pair of wings. From the Middle Kingdom onward, the sun-disc became a widespread symbol of protection.

DABOG/DAZHBOG

A south Slavonic sun-god. Born anew at dawn—young and handsome, aging until death at sunset. In the epic poetry of the Serbians, Dabog appears as ruler over the earth. In the "Song of Igor" we are told that the Russians are Dabog's grandchildren.

HAMMON

The god of the setting sun worshipped by the ancient Lybians. He was depicted with the horns of a ram and is thought to coincide with the oasis god, Ammon.

HARACHTE

Egyptian god of the morning sun. The name means "Horus of the horizon" or "of the place where the sun rises." Harachte was represented as a falcon, and he tended to coalesce with the sun-god Re in the figure of Re-Harachte. In one temple founded by Ramses II he is invoked as "mighty god, lord of heaven."

HELIOS

Greek sun-god, the son of the Titans Hyperion and Theia; brother of the moon-goddess Selene. Helios sees all and hears all, and is invoked as witness to an oath sworn. As god of light, he can make the blind see—but equally, he can strike sinners blind. Helios was often represented driving a chariot drawn by four winged horses, his head surrounded by a halo of rays. In late classical times he was equated with Apollo. His Roman counterpart was Sol.

HORUS

Egyptian god, whose name—meaning "he who is above, he who is afar"—would seem to indicate a sky-god. He was depicted as a falcon, with the sun and the moon as his eyes. At the very beginning of the historical period, the king was equated with this divine falcon. As bearer of the solar eye, Horus is closely connected with the sun-god Re.

HUIRACOCHA

The supreme god of the Inca empire. According to myth he was born of a virgin, and he often displays solar characteristics. His epithet, Pachamac, designates him as "creator of the world." Those who fail to pay fitting homage to the god are destroyed by fire or flood.

HUITZILOPOCHTLI

Tribal god of the Aztecs, associated with the sun at its zenith. His animal manifestation is the hummingbird, which is also the symbol of the sun; his weapon is the "turquoise snake," the symbol of earthly and heavenly fires. He is the adversary of the moon and the stars. Huitzilopochtli embodies the morning sun, the daytime sky, the summer and the south, all of which makes him the luminous adversary of the dark, Tezcatlipoca.

INTI

The Inca sun-god, the object of particular veneration, along with the creator-god, Huiracocha. Inti was seen by the Inca rulers as their progenitor. He was represented by a gold disc with a human face and surrounded by rays.

KINICH KAKMO
The sun-god of the Maya; the
fire-bird corresponding to the
solar aspect of the Aztec
Quetzalcoatl.

LISA/LEZA/LISSA
The name given to the chameleon
sun—worshiped by the Fong of
West Africa; from its union with
the female moon, Mawu, there
arose seven divine pairs of twins.
One of these pairs—the iron
twins—gave mankind the first
tools and weapons.

MAHES
A god in the form of a lion who
was worshipped as a sun-god in
ancient Egypt, especially in the
Nile Delta. As "Lord of
Slaughter," he represents the
consuming powers of the scorching
heat of midsummer.

MARDUK

The great sun-god of Babylon from the time of King Hammurabi onward, the tutelar head god of the Babylonian empire. Considered an agricultural god, the god of wisdom, judgment, and the bringer of light, Marduk was regarded as the "Lord of Kings." He is symbolized by a reptilian dragon, the spade, or the sickle.

MARICI

A Buddhist goddess particularly invoked at sunrise. She is described in Buddhist texts as surrounded by a garland of dazzling rays of light and traveling in a vehicle drawn by seven boars.

MITHRAS

The Graeco-Latin name for the Iranian Mithra, whose cult and mysteries were spread by Roman troops and seafarers over the whole Roman empire during the first and second centuries A.D. The religion of Mithras allowed only men to take part in its mysteries. As the god of loyalty, truth, and the fight against evil, Mithras was a favorite of soldiers. His original identity as a god of light was gradually intensified until he appeared as the figure of Sol invictus, the invincible sun.

RE/RA

The old Egyptian name of the sun and the sun-god. From the fourth dynasty onward the Pharaohs described themselves as sons of Re. In his barque, the sun-god travels across the ocean of heaven, accompanied by his vizier, Thot, and his daughter, Maat. The orb of the sun was taken as the visible body of the god, but it was also regarded as his eye. Symbols of his cult were the obelisks, whose pinnacles were struck by the first rays of the rising sun.

SAMAS/SHAMASH

The Semitic word for the sun became the Babylonian sun-god; during the day he sees all things and hence is the god of justice and of the oracle. Son of the moon-god, Sin, he is symbolized by the sun's disc and a four-pointed star surrounded by rays. In Assyria, the emblem was the winged solar disc. By night, Samas moves through the underworld, bringing nourishment and light to the dead. See Utu.

SAVITAR/SAVITRI

The Hindu solar god who oversees the whole span of heaven and the principle of movement. He drives a golden chariot and has golden arms, eyes, and tongue, which reach the frontiers of heaven.

SHAKURU

The famous Sun Dance of the Pawnees painfully required young men wishing manhood and to overcome human limitations to pierce the flesh on their chest and remain hanging throughout the day following the movement of Shakuru from the place of creation in the east, to the west.

SHEN YI

Chinese sun-god known as the divine archer. Once there were ten suns in heaven whose heat threatened all life on earth. Shen Yi shot down nine of them and became the lord of the remaining one. His consort, Heng E, is goddess of the moon.

SOL

(1) Roman sun-god corresponding to the Greek Helios. His temple stood on the Quirinal, and as patron of the chariot teams, he also had a second temple in the Circus Maximus.

SOL

(2) Old Icelandic name for the sun. Personified as the daughter of Mundilferi (the old moon) and Mani (the new moon). Her cart is drawn across the sky by two horses.

SUAIXTIX

A designation of the sun and the name of the sun-god of the ancient Prussians.

SUL

A Celtic goddess, her name meaning "sun," worshipped in southern England, especially near the area of Bath. An eternal fire burned in her temple. During the Roman occupation she was identified with Minerva.

SURYA

Hindu sun-god whose father was the sky-god Dyaus, or Indra. With dark-red coloring, golden hair, four arms, and three eyes, Surya drives a chariot drawn by a team of four to seven horses; in his hand he holds a lotus flower, often a discus as well. The daughter of the sun-god is also called Surya.

TAWA

The great sun-spirit of the Pueblo Indians of the North American southwest.

TONATIUH

Aztec sun-god, ruler of fate, usually represented with a crown of eagle feathers. He offered hospitality to those who had died in battle and to women who had died in childbirth. Tonatiuh's worshipper's conducted daily sacrifices to quench his thirst for blood.

UTU

Sumerian sun-god, god of fertility, whose main role is that of guardian of justice.

VIVASVAT

Indian god of the sunrise whose name means "he who lights up." In the Rig Veda he is said to bring fire.

WI

*The omniscient sun-god of the
Sioux Indians who protects and
defends those who are brave and
loyal. The buffalo is particularly
associated with Wi and was often
considered a manifestation of the
god himself.*

WIRIUPRANDI

*"Sun-Woman" to Australian
Aborigines.*

WUNEKAU

*A sun-god worshipped by various
tribes of New Guinea. He is
taken to be the creator of all
things, whose name may be
uttered only in the most extreme
reverence. A wind sent by him
can make women pregnant.
A giant snake, Make, is seen
as a special display of his divine
presence.*

YARHIBOL

*Ancient Arabian sun-god, revered
in Palmyra and Dura Europos
along with the sky-god Bel and
the moon-god Aglibol. He was
the god of the holy spring at Efka
and may have been regarded
as an oracle.*

PAGE 8, *Marsilio Ficino*, from COMMENTARY ON PLATO'S SYMPOSIUM OF LOVE, translated by Sears Jayne. Copyright © 1985 by Sears Jayne and published by Spring Publications in Woodstock, CT 06281.

PAGE 9, *Psalm 19*, from A BOOK OF PSALMS by Stephen Mitchell. Copyright © 1993 by Stephen Mitchell. Reprinted by permission of HarperCollins Publishers, Inc.

PAGE 11, *Congregation of Abraxas*, from THE BOOK OF HOURS. Reprinted by permission of the Unitarian Universalist Association.

PAGE 12, *Pharoah Akhenaten*, from ECHOES OF EGYPTIAN VOICES: AN ANTHOLOGY OF ANCIENT EGYPTIAN POETRY, translated by John L. Foster. Copyright © 1992 by John L. Foster. Reprinted by permission of the University of Oklahoma Press.

PAGE 13, *Egyptian Prayer to Sun God*, from MAN IN THE SUN by Jacquetta Hawkes, published by Random House Inc., Copyright © 1962. Reprinted by permission of the Peters Fraser & Dunlop Group, Ltd.

PAGE 15, *Brihadaranyaka Upanishad, III. vii. 9*, from THE UPANISHADS, VOLUME III, as translated into English by Swami Nikhilananda and published by the Ramakrishna-Vivekananda Center of New York, Copyright © 1956 by Swami Nikhilananda.

PAGE 16, *Saint Francis of Assisi*, from *"Canticle of Brother Sun,"* THE WRITINGS OF SAINT FRANCIS OF ASSISI, published by Ditchling Press, Ltd. for Burns & Oates, Ltd. 1964.

PAGE 17, *Celtic Prayer*, from CELTIC BLESSINGS edited by Caitlín Matthews. Copyright © 1994 Element, Inc., Rockport, MA.

PAGE 23, *Rumi*, from OPEN SECRET: VERSIONS OF RUMI, translated by John Moyne and Coleman Barks. Copyright © 1984 by Threshold Books, RD4, Box 600, Putney, VT, 05346.

PAGE 25, *Oglala Sioux*, from THE SACRED PIPE: BLACK ELK'S ACCOUNT OF THE SEVEN RITES OF THE OGLALA SIOUX, recorded and edited by Joseph Epes Brown. Copyright © 1953, 1989 by the University of Oklahoma Press.

PAGE 27, *Sappho of Lesbos*, from *"Standing By My Bed,"* from SAPPHO: A NEW TRANSLATION, translated by Mary Barnard. Copyright © 1958 by the Regents of the University of California; © renewed 1984 Mary Barnard. Reprinted by permission of the University of California Press.

ACKNOWLEDGMENTS

PAGE 49, *Doris Lessing*, from BRIEFING FOR A DESCENT INTO HELL. Copyright © 1971 by Doris Lessing. First Vintage Books Edition, May 1981. Originally published by Alfred A. Knopf, Inc.

PAGES 50-51, *Artur Lundkvist*, from JOURNEYS IN DREAM AND IMAGINATION, translated by Ann B. Weissmann and Annika Planck. Published by Four Walls Eight Windows. Reprinted by permission of the publisher.

PAGE 53, *Martha Heyneman*, reprinted from PARABOLA, *The Magazine of Myth and Tradition*, VOL. VIII, NO. 4, Winter, 1983.

PAGE 59, *Rig Veda*, from THE RIG VEDA, selected and translated by Wendy Doniger O'Flaherty from "*The Sun, Surya,*" Copyright © 1981 by Wendy Doniger O'Flaherty. Reprinted by permission of Penguin Books, Ltd. (Penguin Classics, 1951).

PAGE 60, *Theodore Roethke*, from "*What Can I Tell My Bones?*" from THE COLLECTED POEMS OF THEODORE ROETHKE. Copyright © 1957, used by permission of Doubleday, a Division of Bantam Doubleday Dell Publishing Group, Inc., and Faber & Faber Ltd., UK.

PAGE 65, *St. Symeon*, from THREE BYZANTINE SACRED POETS, edited by N. M. Vaporis. Copyright © 1979 by Hellenic College Press, Brookline, MA.

PAGE 67, *Dylan Thomas*, from POEMS OF DYLAN THOMAS. Copyright © 1952 by the Trustees for the Copyrights of Dylan Thomas. Reprinted by permission of New Directions Publishing Corp., and David Higham Associates Ltd., UK.

*An extremely thorough research effort has been made to locate
all rights holders and to obtain appropriate reprint permissions.
If any required acknowledgments have been omitted, or any rights overlooked,
it is unintentional. If notified, the editor of this book will rectify
any omission in future editions.*